Oksan Oral

Effects of Model Properties on Fabric Usage Amount and Costs of Labour

AF153244

Oksan Oral

Effects of Model Properties on Fabric Usage Amount and Costs of Labour

Model Characteristics, Cycle Time, Methodology

LAP LAMBERT Academic Publishing

Impressum / Imprint

Bibliografische Information der Deutschen Nationalbibliothek: Die Deutsche Nationalbibliothek verzeichnet diese Publikation in der Deutschen Nationalbibliografie; detaillierte bibliografische Daten sind im Internet über http://dnb.d-nb.de abrufbar.
Alle in diesem Buch genannten Marken und Produktnamen unterliegen warenzeichen-, marken- oder patentrechtlichem Schutz bzw. sind Warenzeichen oder eingetragene Warenzeichen der jeweiligen Inhaber. Die Wiedergabe von Marken, Produktnamen, Gebrauchsnamen, Handelsnamen, Warenbezeichnungen u.s.w. in diesem Werk berechtigt auch ohne besondere Kennzeichnung nicht zu der Annahme, dass solche Namen im Sinne der Warenzeichen- und Markenschutzgesetzgebung als frei zu betrachten wären und daher von jedermann benutzt werden dürften.

Bibliographic information published by the Deutsche Nationalbibliothek: The Deutsche Nationalbibliothek lists this publication in the Deutsche Nationalbibliografie; detailed bibliographic data are available in the Internet at http://dnb.d-nb.de.
Any brand names and product names mentioned in this book are subject to trademark, brand or patent protection and are trademarks or registered trademarks of their respective holders. The use of brand names, product names, common names, trade names, product descriptions etc. even without a particular marking in this works is in no way to be construed to mean that such names may be regarded as unrestricted in respect of trademark and brand protection legislation and could thus be used by anyone.

Coverbild / Cover image: www.ingimage.com

Verlag / Publisher:
LAP LAMBERT Academic Publishing
ist ein Imprint der / is a trademark of
OmniScriptum GmbH & Co. KG
Heinrich-Böcking-Str. 6-8, 66121 Saarbrücken, Deutschland / Germany
Email: info@lap-publishing.com

Herstellung: siehe letzte Seite /
Printed at: see last page
ISBN: 978-3-659-50339-9

Zugl. / Approved by: Izmir, Ege University, Diss., 2003

TABLE OF CONTENTS   **Page**

I

TABLE OF CONTENTS   Page

To my mother and my father

ACKNOWLEDGEMENTS

I wish to thank my supervisor, Prof. Dr. Mehmet Cetin Erdogan, for his valuable academic support. And, I thank my family especially my mother and my father for their efforts in my academic life.

Three scientific papers were published in this PhD Thesis. These were:
Oral, O., Erdogan, M. C., Dirgar, E., 2013. The Relationship Between Model Types and Related Parameters, Industria Textila, Vol: 64,   Issue: 4   ISSN: 1222-5347, p. 210-216

Kansoy, O., Erdogan, M. C., 2006, The Relationship Between Model Properties and Number of Pieces, Perimeter of Pieces and Sewing Time, Tekstil ve Konfeksiyon, Vol:16 , Issue: 1, İzmir, p.320-324.

Kansoy, O., Erdogan, M. C., Ondogan Aktuglu, Z., 2005, The Effects of Model Properties on Cutting Time, Tekstil ve Konfeksiyon, Vol:15 , Issue: 3, İzmir, p.172-177.

1. INTRODUCTION

One of the most important key factors of increasing compatibility in textile is being able to reduce the costs to global averages. The compatibility of apparel manufacturers depends on product standardization; technology; how advanced they are and whether they have the mental work-power to deal with the technology. The technologic development is not limited only with automation of the machinery on the production line but also includes processes before and after the production.

In companies, technologic improvements are towards increasing productivity and reducing costs. Production is vitally important for the companies. Improvement of productivity is not only increasing profit but also the improvement in the way of production. Material and work-power are the basic subjects of the efforts on improving productivity. The benefits depending on the structures of the products models, will directly affect the productivity of the company.

In companies, cost determination is the primary issue, on which most attention and care is paid. Since all savings without sacrificing from the quality affect the costs in a positive manner, savings from the materials and production time should be the primary target. One of the main factors that affect the cost of a product is the characteristics of the model. It should be taken into consideration that model's being in fewer numbers of pieces will affect the amount of fabric to be used; its cutting and sewing times.

It is obvious that customer demands directed the manufacturer to work with different models. Various models can be formed for garment types. Forming a model on a garment may be defined as; cutting the garment into desired pieces; chancing the form of the garment by dividing it into pieces. When forming a model, it is important to consider all phases of the production line and doing it economically with existing resources.

In the study, the effects of different model characteristics of garments on fabric usage amount and cutting and sewing times have been investigated. Following the experiment, a method for determining cutting and sewing times for sample product groups has been created.

In apparel companies, after the model is formed, this method will be useful in; estimating cutting and sewing times; calculating the productivity; production planning and estimating delivery date.

2. GENERAL INFORMATION

2.1. The Work Flow in Apparel Companies

The apparel industry is an industry which needs a lot of labor; therefore the production process requires a good level of technical knowledge together with extensive and intricate work. By using high-technology, it is possible to do the production in a shorter time, with fewer mistakes; it can become more productive and be of good quality.

In the apparel companies, the first link of the chain of production is preparing clothing patterns. After the clothing pattern belonging to the base size is prepared, in order to make the patterns of the other sizes in accordance with the given measurement table, the grading process of the base size is carried out. After this stage, the marker plan preparations begin. Marker plan preparation is the placement of the specified model – in accordance with the specified number of sizes – on an image of fabric of clothing patterns with a certain width by using the fabric as economical as possible. Pattern preparation, grading and marker plan preparation can be done by two different methods.

- Preparing the patterns, grades of patterns and marker plans manually.

- Preparing the patterns, grades of patterns and marker plans with Computer Aided Design Systems (CAD Systems).

The second stage of production is executed in the cutting department. The fabric laying process is done according to the marker plan prepared in this department. Laying can be done manually or by automatic spreader machines. After the laying process is over, the cutting process may begin. The cutting process can be done by two methods:

- Manual cutting (with straight knife, round knife and band knife)

- Cutting with Numeric Cutting Systems (CNC-Cutter).

Another important sector of production is the sewing department. The production in the sewing department is organized in two stages: preparation and fitting. The major part of the total production time of an apparel product is spent in the sewing processes. In this department, each piece which forms a product and all the processes that are done on these pieces are very intricate and intense.

The pieces which come out of the sewing room are sent to quality control, ironing and packing departments in sequence and they are taken to the warehouse where they wait for the delivery time of the manufactured goods.

2.2. Technology and Automation at The Production Stages in The Apparel Industry

The increase of production productivity involves determining the main factors of the production system and being able to apply them. The technology factor, which is one of the factors which affect the increase of productivity, is an important detractive power in a competitive environment. The apparel industry still continues to be a labor dominated industry in a competitive environment. However, the increase of the model variety due to the fashion and demogratical factors has brought specializing with an inclination to have a variety in the production of products.

The customers are not satisfied with classical products, ones which are produced with general intentions and ones which don't have many alternatives and they demand to have more models to choose from. It is difficult to meet these demands with traditional design and production methods. For this reason high technology is used a great deal in the stages of design and production. The most important systems used for providing a high technology support in the apparel industry are Computer Aided Design-CAD and Computer Aided Manufacturing-CAM systems.

2.2.1. CAD System

CAD means designing a product with the help of a computer. Product design is the main factor that affects the quality and the cost of a product. Many manufacturers prefer CAD systems. CAD systems are faster, more consistent, accurate and manageable than manual methods.

In the computer-based pattern preparation and grading system, production patterns can be transferred to a computer by two methods.

- Preparing the base size pattern manually and transferring it to the computer by defining it on the digitizer.

- Preparing the base size pattern on the computer with the "pattern preparation from scratch" methods.

2.2.2. CAM System

CAM means a production, supported by a computer, in order to decrease the person made mistakes. In the apparel industry, the cutting department is the department where the CAM system is used most frequently. The correct and productive works carried out in the cutting department, has a positive effect on the quality and the productivity of a company. Cutter saves time and saves from important cost factors such as fabric and labor. It also helps increase the cutting quality and productivity.

While working on a process with an automatic cutter, the marker plans need to be prepared on the computerized pattern preparation and grading system and need to be sent to the cutter. The cutter carries out the cutting process in line with the marker plan in its memory.

2.3. Cost in Apparel Company

When the cost of the apparel products are studied, the factors that form the cost and their values are as follows :

Raw material and accessories = 50-60 %

Labor   = 20-25 %

General expenditures   = 20-25 %

In order to lower the cost of fabric, it is required to decrease the amount of fabric usage. This can be possible by preparing the patterns and grading with CAD systems and preparing the marker plans so as to make the least fabric wastage.

2.4. Method-Time Measurement (MTM)

Method's Time Measurement (MTM) is a predetermined motion time system that is used primarily in industrial settings to analyze the methods used to perform any manual operation or task and, as a product of that analysis, set the standard time in which a worker should complete that task.

It is a method which separates the flow of motion from basic motions. There are some certain basic motions which are used to perform an operation. The factors, that affect this motion, have been analyzed, the time consumed in order to perform this motion has been determined as "norm time" and "norm time tables" have been prepared.

The MTM method has been applied according to the stages below:

- Operation analysis,

- Estimation of effective time,

- Taking the effective time from the norm time tables,

- Adding up all the effective times which form the operation and determining the standard unit time for that operation.

Norm time tables are international tables and the unit time is valid all over the world. Through the using MTM, method of operation is determined in detail. In this method, there is no need to estimate performance. Time value is defined according to normal performance (100% performance). MTM is used to determine time, method of operation, job training, and pricing.

In the apparel industry, during the stage where the MTM method is used, two norm table values are used.

4

• Hand motion table
• Stitch length table

By taking into account that two table value, unit time is calculated.

3. MATERIAL AND METHOD

The material of the study consisted of clothing product groups made of woven fabric and garment models, CAD (computer aided design) machine used in model pattern department and CAM (computer aided manufacturing) cutting machine used in cutting department.

Product groups subject to the study are skirt, lady's trousers, dress and men's coat. In forming the models to be applied to each product group, some criteria were taken into consideration. In order to be able to compare different modeling within each group, various dividing were made and the criteria for models were determined. These are; classic, horizontally cut, vertically cut and both vertically and horizontally cut types of models. In cutting of models some drawing bases were benefited from. Since investigation of the relationship of the number of pieces with fabric usage amount and cutting and sewing time was one of the aims of the study, preparing models with higher number of pieces was a priority.

In this study, "Evaluation research method" was chosen as the evaluation method. The method includes an analytic evaluation suitable for the aim of the study, conduction of the study experiments and data analyses.

Analysis of production process and production flow:

Step 1- Preparations of the patterns of the chosen models using Muller pattern system according to CAD system

Step 2- Grading of patterns according to determined sizes

Step 3- Preparation of marker plans for models

Step 4- Cutting of marker plans using CAM

Step 5- Determination of sewing times for the models using the method MTM

Pre accepted properties of the study

Fabric:

The woven fabric used in the study is single colored; has no nap and available for placing the patterns in both directions. The 148 cm wide fabric (the most common one in the market) was used and all pattern placements were practiced on 148 cm width.

Sizes:

Data for normal men and women size table which are used in preparing the patterns are given in Appendix 1 and Appendix 2.

Assortment of sizes:

Before the patterns were prepared, sizes to be used in the experiment and their numbers together with their assortment numbers have to be determined. In this study, sizes and the assortments used for the woman group are shown below:

<u>36</u>   <u>38</u>   <u>40</u>   <u>42</u>
1   1   1   1

Following sizes and assortments are for the man group:

<u>46</u>   <u>48</u>   <u>50</u>   <u>52</u>
1   1   1   1

Determination of number of samplings:

In the study, 8 different models to the chosen skirt, lady's trousers and dress group and 7 different model to the men's coat group were applied.

The experimental pattern of the study:

In the study the following parameters were investigated.

Total number of pieces: Total numbers of pieces on the cutting plan belonging to sizes were taken into account.

Fabric usage proportion: The marker plans of the sizes of the model were prepared with the CAD System and the value of this placement was taken as a percentage.

Area of pieces: Total value of the area of the sizes on the marker plan was taken in m^2.

Perimeter of the pieces: Total length of the perimeters of the sizes on the marker plan was taken in meters.

Cutting time: The duration of the CAM operated cutting process of a single layer of the cutting plan was taken in minutes.

Sewing time: Taking the production diagrams of the models into consideration, operation unit times were taken in minutes according to MTM (Method's time Measurement).

The results of the trials were evaluated using "SPSS" (statistical packet software).

"Pearson correlation" was used in the study. If "p" the "significance value" is smaller than 0.05 (probability) the linear relation (positive correlation) ($p < 0.05$)

between variables is significant. And if p>0.05, there is no positive correlation, therefore, insignificant.

Regression analysis was conducted to investigate the effects of other happenings on the observed process.

Data for all of values are given in Appendix 34, 35, 36, 37.

4. FINDINGS

4.1. Findings For Fabric Usage Proportion, Total Number of Pieces, Area of Pieces, and Perimeter of Pieces on The Marker

4.1.1. Skirt group

Correlation analyses of the results of trials conducted as 8 trials for skirt models were investigated using the software "SPSS". Values obtained as a result of correlation are given in Table 1.

Table 1. Correlation for the fabric usage proportion, the total number of pieces, the area of the pieces, and the perimeter of pieces on the marker for the skirt group.

Factors	r	p	n
Fabric usage proportion – Total number of pieces	0.489	0.219	8
Fabric usage proportion – Area of pieces	-0.14	0.974	8
Total number of pieces – Perimeter of pieces	0.972	0	8
Total number of pieces – Area of pieces	-0.332	0.421	8

When the results in Table 1 is examined;
- Since the relationship between fabric usage proportion and total number of pieces is p=0.219, the linear relationship between these variables is statistically insignificant.
- Since the relationship between fabric usage proportion and area of pieces is p=0.974, the linear relationship between these variables is statistically insignificant.
- Since the relationship between total number of pieces and perimeter of the pieces is p=0, the linear relationship between these variables is statistically significant.
- Since the relationship between total number of the pieces and area of pieces is p=0.421, the linear relationship between these variables is statistically insignificant.

Figure 1 shows the changing in the perimeter of pieces depending on the number of pieces.

7

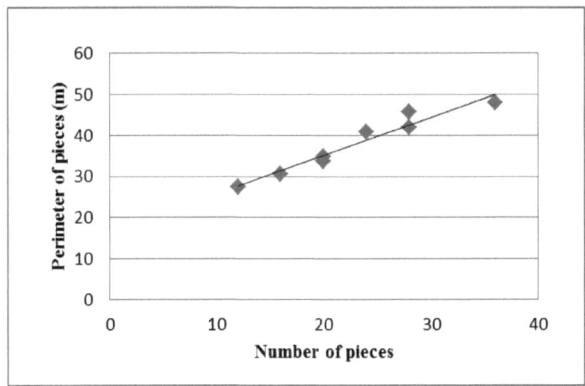

Figure 1. Changes in the perimeter of pieces depending on the number of pieces in skirt models.

4.1.2. Lady's trousers group

Correlation analyses of the results of trials conducted as 8 trials for lady's trousers models were investigated using the software "SPSS". Values obtained as a result of correlation are given in Table 2.

Table 2. Correlation for the fabric usage proportion, the total number of pieces, the area of the pieces, and the perimeter of pieces on the marker for the lady's tousers group

Factors	r	p	n
Fabric usage proportion – Total number of pieces	0.566	0.144	8
Fabric usage proportion – Area of pieces	0.591	0.123	8
Total number of pieces – Perimeter of pieces	0.832	0.01	8
Total number of pieces – Area of pieces	-0.157	0.711	8

When the results in Table 2 is examined;
- Since the relationship between fabric usage proportion and total number of pieces is p=0.144, the linear relationship between these variables is statistically insignificant.
- Since the relationship between fabric usage proportion and area of pieces is p=0.123, the linear relationship between these variables is statistically insignificant.
- Since the relationship between total number of pieces and perimeter of the pieces is p=0.01, the linear relationship between these variables is statistically significant.
- Since the relationship between total number of the pieces and area of pieces is p=0.711, the linear relationship between these variables is statistically insignificant.

Figure 2 shows the changing in the perimeter of pieces depending on the number of pieces.

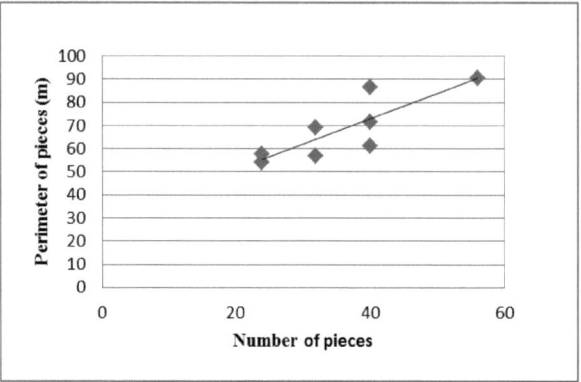

Figure 2. Changes in the perimeter of pieces depending on the number of pieces in lady's trousers models.

4.1.3. Dress group

Correlation analyses of the results of trials conducted as 8 trials for dress models were investigated using the software "SPSS". Values obtained as a result of correlation are given in Table 3.

Table 3. Correlation for the fabric usage proportion, the total number of pieces, the area of the pieces, and the perimeter of pieces on the marker for the dress group

Factors	r	p	n
Fabric usage proportion – Total number of pieces	0.069	0.870	8
Fabric usage proportion – Area of pieces	0.065	0.878	8
Total number of pieces – Perimeter of pieces	0.887	0.003	8
Total number of pieces – Area of pieces	-0.274	0.511	8

When the results in Table 3 is examined;
- Since the relationship between fabric usage proportion and total number of pieces is p=0.870, the linear relationship between these variables is statistically insignificant.
- Since the relationship between fabric usage proportion and area of pieces is p=0.878, the linear relationship between these variables is statistically insignificant.
- Since the relationship between total number of pieces and perimeter of the pieces is p=0.003, the linear relationship between these variables is statistically significant.
- Since the relationship between total number of the pieces and area of pieces is p=0.511, the linear relationship between these variables is statistically insignificant.

9

Figure 3 shows the changing in the perimeter of pieces depending on the number of pieces.

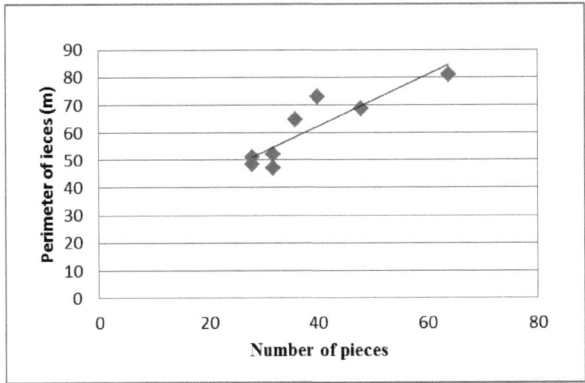

Figure 3. Changes in perimeter of pieces depending on the number of pieces in dress models.

4.1.4. Men's coat group

Correlation analyses of the results of trials conducted as 7 trials for men's coat models were investigated using the software "SPSS". Values obtained as a result of correlation are given in Table 4.

Table 4. Correlation for the fabric usage proportion, the total number of pieces, the area of the pieces, and the perimeter of pieces on the marker for the men's coat group

Factors	r	p	n
Fabric usage proportion – Total number of pieces	0.822	0.023	7
Fabric usage proportion – Area of pieces	0.951	0.001	7
Total number of pieces – Perimeter of pieces	0.914	0.004	7
Total number of pieces – Area of pieces	0.912	0.004	7

When the results in Table 4 is examined;

- Since the relationship between fabric usage proportion and total number of pieces is p=0.023, the linear relationship between these variables is statistically significant.
- Since the relationship between fabric usage proportion and area of pieces is p=0.001, the linear relationship between these variables is statistically significant.
- Since the relationship between total number of pieces and perimeter of the pieces is p=0.004, the linear relationship between these variables is statistically significant.
- Since the relationship between total number of the pieces and area of pieces is p=0.004, the linear relationship between these variables is statistically significant.

Figure 4 shows the changing in the perimeter of pieces depending on the number of pieces.

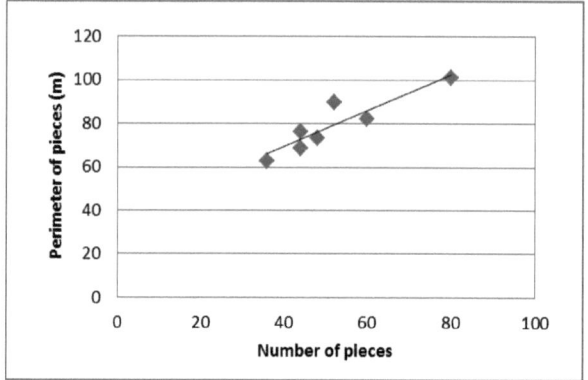

Figure 4. Changes in the perimeter of pieces depending on the number of pieces in men's coat models.

Figure 5 shows the changing in the fabric usage proportion depending on the number of pieces.

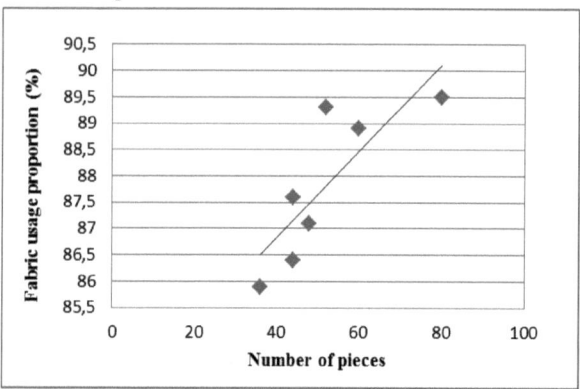

Figure 5. Changes in the fabric usage proportion depending on the number of pieces in men's coat models.

Figure 6 shows the changing in the fabric usage proportion depending on the area of pieces.

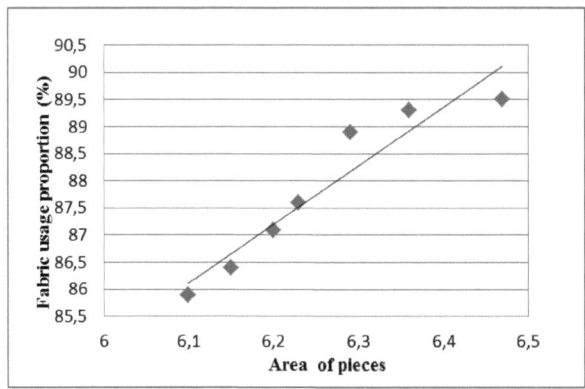

Figure 6. Changes in the fabric usage proportion depending on the area of pieces in men's coat models.

4.2. Findings Regarding Number of Pieces, Cutting Time and Sewing Time

4.2.1. Skirt group

Correlation analyses of the results of trials conducted as 8 trials for skirt models were investigated using the software "SPSS". Values obtained as a result of correlation are given in Table 5.

Table 5. The correlation of number of pieces, cutting time and sewing time in skirt groups

Factors	r	p	n
Number of pieces - cutting time	0.549	0.159	8
Number of pieces - sewing time	0.968	0	8

When the results in Table 5 are examined;
- Since the relationship between number of pieces and cutting time is p=0.159, the linear relationship between these variables is statistically insignificant.
- Since the relationship between the number of pieces and sewing time is p=0, the linear relationship between these variables is statistically significant.

Figure 7 shows the changing in the sewing time depending on the number of pieces.

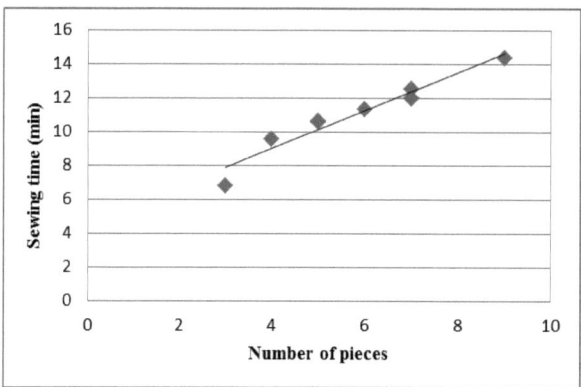

Figure 7. Changes in the sewing time depending on the number of pieces in skirt models.

4.2.2. Lady's trousers group

Correlation analyses of the results of trials conducted as 8 trials for lady's trousers models were investigated using the software "SPSS". Values obtained as a result of correlation are given in Table 6.

Table 6 The correlation of number of pieces, cutting time and sewing time in lady's trousers groups

Factors	r	p	n
Number of pieces - cutting time	0.813	0.014	8
Number of pieces - sewing time	0.942	0	8

When the results in Table 6 are examined;
- Since the relationship between number of pieces and cutting time is p=0.014, the linear relationship between these variables is statistically significant.
- Since the relationship between the number of pieces and sewing time is p=0, the linear relationship between these variables is statistically significant.

Figure 8 shows the changing in the cutting time depending on the number of pieces

13

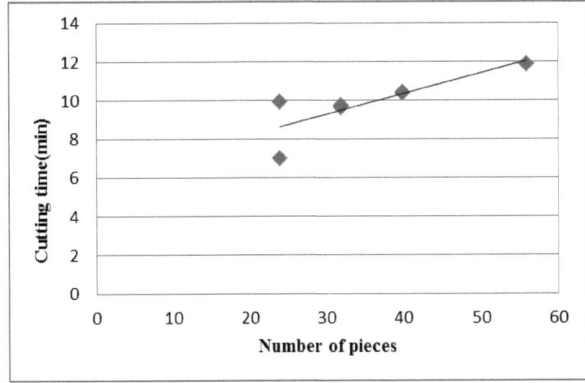

Figure 8. Changes in the cutting time depending on the number of pieces in lady's trousers models.

Figure 9 shows the changing in the sewing time depending on the number of pieces

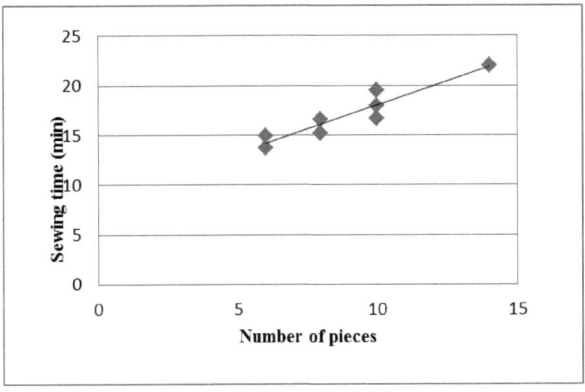

Figure 9. Changes in the sewing time depending on the number of pieces in lady's trousers models.

4.2.3. Dress group

Correlation analyses of the results of trials conducted as 8 trials for dress models were investigated using the software "SPSS". Values obtained as a result of correlation are given in Table 7.

Table 7 The correlation of number of pieces, cutting time and sewing time in dress groups

Factors	r	p	n
Number of pieces - cutting time	0.634	0.092	8
Number of pieces - sewing time	0.936	0.001	8

When the results in Table 7are examined;
- Since the relationship between number of pieces and cutting time is p=0.092, the linear relationship between these variables is statistically insignificant.
- Since the relationship between the number of pieces and sewing time is p=0.001, the linear relationship between these variables is statistically significant.

Figure 10 shows the changing in the sewing time depending on the number of pieces

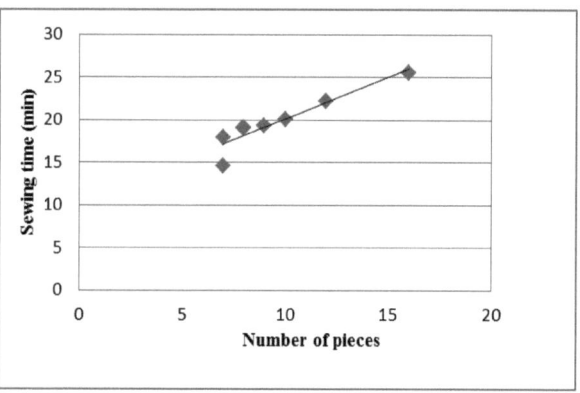

Figure 10. Changes in the sewing time depending on the number of pieces in dress models.

4.2.4. Men's coat group

Correlation analyses of the results of trials conducted as 7 trials for men's coat models were investigated using the software "SPSS". Values obtained as a result of correlation are given in Table 8.

Table 8. The correlation of number of pieces, cutting time and sewing time in men's coat groups

Factors	r	p	n
Number of pieces - cutting time	0.981	0	7
Number of pieces - sewing time	0.942	0.002	7

When the results in Table 8 are examined;
- Since the relationship between number of pieces and cutting time is p=0, the linear relationship between these variables is statistically significant.
- Since the relationship between the number of pieces and sewing time is p=0.002, the linear relationship between these variables is statistically significant.

Figure 11 shows the changing in the cutting time depending on the number of pieces.

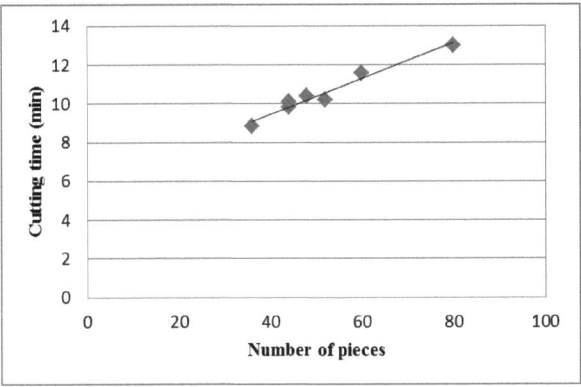

Figure 11. Changes in the cutting time depending on the number of pieces in men's coat models.

Figure 12 shows the changing in the sewing time depending on the number of pieces.

16

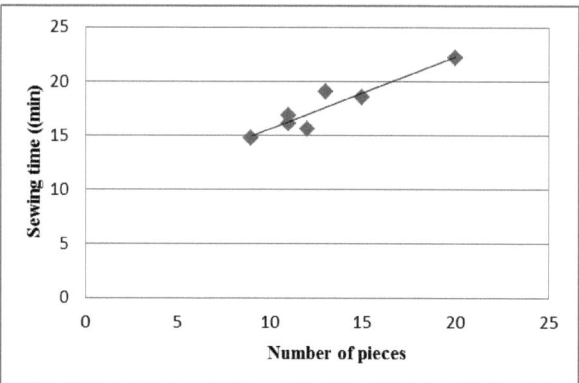

Figure 12. Changes in the sewing time depending on the number of pieces in men's coat models

4.3. Findings Regarding Perimeter of Pieces, Cutting Time and Sewing Time
4.3.1. Skirt group

Correlation analyses of the results of trials conducted as 8 trials for skirt models were investigated using the software "SPSS". Values obtained as a result of correlation are given in Table 9.

Table 9. The correlation of perimeter of pieces, cutting time and sewing time in skirt groups

Factors	r	p	n
Perimeter of pieces - cutting time	0.573	0.137	8
Perimeter of pieces - sewing time	0.929	0.001	8

When the results in Table 9 are examined;
- Since the relationship between perimeter of pieces and cutting time is p=0.137, the linear relationship between these variables is statistically insignificant.
- Since the relationship between the perimeter of pieces and sewing time is p=0.001, the linear relationship between these variables is statistically significant.

Figure 13 shows the changing in the sewing time depending on the perimeter of pieces.

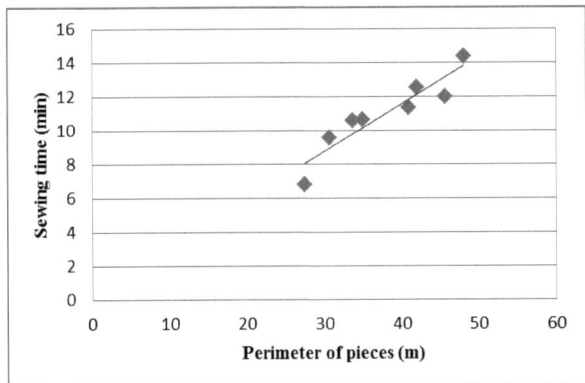

Figure 13. Changes in the sewing time depending on the perimeter of pieces in skirt models.

4.3.2. Lady's trousers group

Correlation analyses of the results of trials conducted as 8 trials for lady's trousers models were investigated using the software "SPSS". Values obtained as a result of correlation are given in Table 10.

Table 10. The correlation of perimeter of pieces, cutting time and sewing time in lady's trousers groups

Factors	r	p	n
Perimeter of pieces - cutting time	0.720	0.044	8
Perimeter of pieces - sewing time	0.946	0	8

When the results in Table 10 are examined;

- Since the relationship between perimeter of pieces and cutting time is p=0.044, the linear relationship between these variables is statistically significant.

- Since the relationship between the perimeter of pieces and sewing time is p=0, the linear relationship between these variables is statistically significant.

Figure 14 shows the changing in the cutting time depending on the perimeter of pieces.

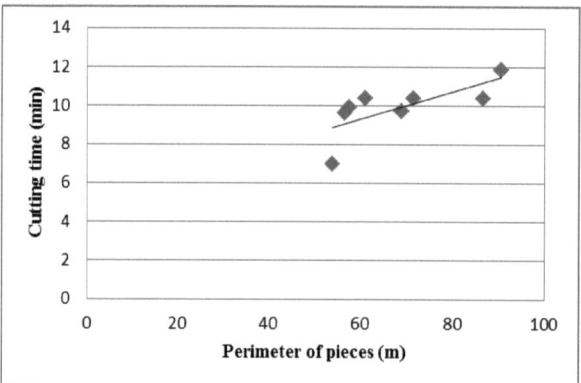

Figure 14. Changes in the cutting time depending on the perimeter of pieces in lady's trousers models.

Figure 15 shows the changing in the sewing time depending on the perimeter of pieces.

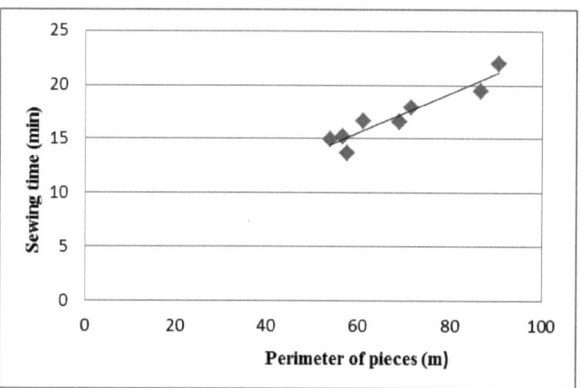

Figure 15. Changes in the sewing time depending on the perimeter of pieces in lady's trousers models.

4.3.3. Dress group

Correlation analyses of the results of trials conducted as 8 trials for dress models were investigated using the software "SPSS". Values obtained as a result of correlation are given in Table 11.

Table 11. The correlation of perimeter of pieces, cutting time and sewing time in dress groups.

Factors	r	p	n
Perimeter of pieces - cutting time	0.593	0.121	8
Perimeter of pieces - sewing time	0.793	0.019	8

When the results in Table 11 are examined;

-Since the relationship between perimeter of pieces and cutting time is p=0.121, the linear relationship between these variables is statistically insignificant.

- Since the relationship between the perimeter of pieces and sewing time is p=0.019, the linear relationship between these variables is statistically significant.

Figure 16 shows the changing in the sewing time depending on the perimeter of pieces.

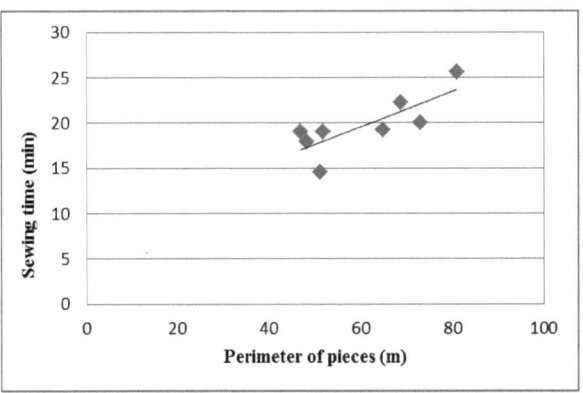

Figure 16. Changes in the sewing time depending on the number of pieces in dress models.

4.3.4. Men's coat group

Correlation analyses of the results of trials conducted as 7 trials for men's coat models were investigated using the software "SPSS". Values obtained as a result of correlation are given in Table 12.

Table 12. The correlation of perimeter of pieces, cutting time and sewing time in men's coat groups.

Factors	r	p	n
Perimeter of pieces - cutting time	0.866	0.012	7
Perimeter of pieces - sewing time	0.977	0	7

When the results in Table 12 are examined;

-Since the relationship between perimeter of pieces and cutting time is p=0.012, the linear relationship between these variables is statistically significant.

- Since the relationship between the perimeter of pieces and sewing time is p=0, the linear relationship between these variables is statistically significant.

Figure 17 shows the changing in the cutting time depending on the perimeter of pieces.

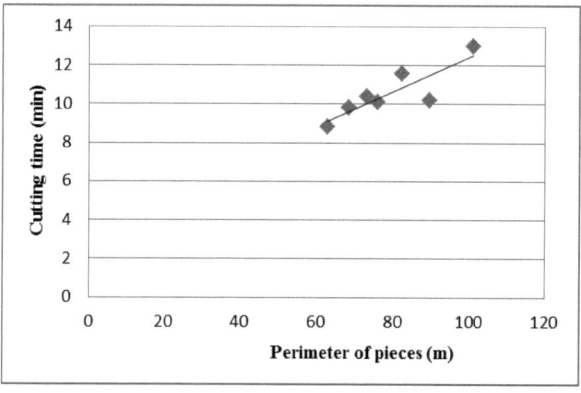

Figure 17. Changes in the cutting time depending on the perimeter of pieces in men's coat models.

Figure 18 shows the changing in the sewing time depending on the perimeter of pieces.

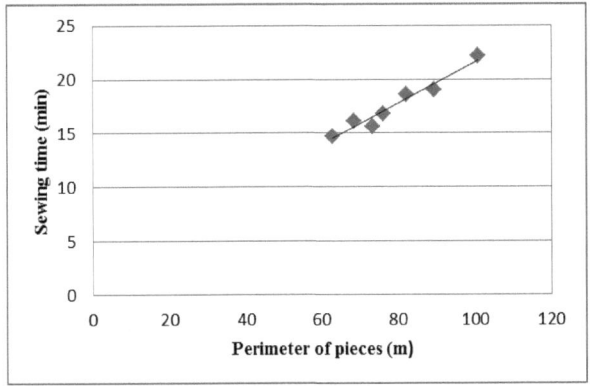

Figure 18. Changes in the sewing time depending on the perimeter of pieces in men's coat models

5. RESULTS

5.1. Results Regarding Fabric Usage Proportion, Total Number of Pieces on The marker, Area of Pieces and Perimeter of Pieces

During the studies made on fabric usage proportion, the total number of pieces on the marker, the area of pieces, and the perimeter of pieces, the conclusions below have been reached:

- Results regarding the fabric usage proportion and the total number of pieces on a marker

For skirt, lady's trousers and dress models, the increase of the number of pieces does not have an effect on the fabric usage proportion. There are two important factors which made way for this conclusion:

The first one is; flared skirts, flaring trousers and dress models which have fewer pieces but have a larger pattern area and which have been evaluated with the model feature criteria. For these models, while working with a suitable fabric width, a very tidy placement is formed, the pattern pieces are well placed on the open spaces left from each other on the marker plan and fabric wastage decreases.

The second important factor is that the horizontally cut models are placed on the marker more productively than the vertically cut models. Even though the vertically cut models chosen in this research have many pieces, the placement productivity is less than that of horizontally cut models. In horizontally cut models, the height of each pattern piece is short. Therefore when the placement is being done, more productive placement possibilities which can give a shorter marker height are

22

executed. Whereas in the vertically cut models, it is the height of each pattern that affects the marker height. In such situations, productive results can only happen when suitable fabric width and suitable size distribution is used.

Another important result concluded from all three clothing groups is as follows: both vertically and horizontally cut models are placed on the markers with high productivity.

Two models have been worked with in criteria of horizontally cut and vertically cut models (Appendix 3-33) . The number of pieces is higher for the second models than for the first models. The fabric usage proportion has been found more economical in the second models. Hence, in both horizontally and vertically cut models the fabric usage productivity increases parallel to the increase of numbers of pieces.

In men's coat models, the fabric usage proportion increases as the number of pieces increase. In men's coat models, it is not very practical to work with a flaring model. Therefore 7 models have been applied on men's coats. When the horizontally and vertically cut models are studied, it is seen that there is not much difference concerning the placement productivity. This is because no modeling is done on sleeve and collar pieces but it is only applied on the body part. A difference of approximately   1,5% between the horizontally and vertically cut models is a consequent of the model characteristics being applied on body parts.

As a result, as a general explanation and an explanation for product groups, for the models with higher numbers of pieces, the placement proportion of patterns on fabric, meaning fabric usage productivity, is higher.

From this conclusion, according to the criteria of the width of the fabric, size distribution and model forming, the increase in numbers of pieces increases the fabric usage productivity.

  - Results regarding the fabric usage proportion and the area of pieces

For skirt, trousers and dress models, it is seen that the increase and decrease of the area of pieces does not have an effect on the fabric usage proportion. The factors which helped for this result to be discovered are the same ones which affect the fabric usage proportion and the total number of pieces.

Whereas in men's coat models, as the area of pieces increases the fabric usage proportion increases. One of the strongest reasons for this is that the flaring model is not used for men's coats.

The decrease and increase of the area of pieces in accordance with the models, is formed by the decrease and increase in the seam allowance depending on the number

23

of pieces. As a result of the research, it has been seen that a larger seam allowance is required in vertical modeling than in horizontal modeling. However the increase and decrease of the area depending only on seam allowance does not affect the fabric usage proportion to increase or decrease, but the characteristics of the model and the width of the fabric affect it as well. From these conclusions, it is seen that there is a direct relation between fabric usage proportion and area of pieces for men's coat models.

- Results regarding total number of pieces and perimeter of pieces;

For all the product groups of the study, it was observed that an increase in the number of pieces increases the perimeter of pieces. If the model is divided into higher number of pieces, lines to be attached are prolonged. And this causes the perimeter of pieces that form the model to prolong. According to the results of the study, it was proven that vertically cut models have longer perimeters than horizontally cut models. If the perimeter length is desired to be prolonged, the model has to be divided into higher number of pieces. This increases the number of pieces.

As a result of all the factors and degrees of effect being studied, it is concluded that the characteristics of models are very important factors which affect the fabric usage proportions.

(Data for fabric usage proportion, total number of pieces on the marker, area of pieces and perimeter of pieces are given in Appendix 34).

5.2. Results Regarding Number of Pieces, Cutting Time and Sewing Time

When number of pieces and cutting times were examined;

While the increase in the number of pieces had no effect on cutting time in skirt models and dress models, higher number of pieces increased cutting time in lady's trousers models and men's coat models.

When number of pieces and sewing times are examined;

An increase in the number of pieces in all product groups prolonged the sewing time.

In both cutting and sewing processes another factor related to the number of pieces is the perimeter of pieces. Therefore, it was found more appropriate to evaluate cutting and sewing time data in comparison with the data for number of pieces and perimeter of pieces.

(Data for number of pieces and cutting time are given in Appendix 35, data for number of pieces and sewing time are given in Appendix 36).

5.3. Results Regarding the Perimeter of Pieces, Cutting Time and Sewing Time

In the previous evaluations, a positive correlation was found. An increment in the number of pieces increased the perimeter of pieces. Cutting and sewing processes depend completely on the lengths of perimeters. Higher numbers of pattern pieces on the model will certainly increase the total perimeter lengths and therefore, changes will occur in cutting and sewing times depending on the properties of the systems used in the production.

When number of pieces, perimeter of pieces and cutting times were examined;

It was observed that increments in number of pieces and perimeter of pieces had no effect on cutting time in skirt models and dress models, whereas in lady's trousers models and men's coat models the increments in these elements prolonged the cutting time.

In computer aided cutting machines, cutting times and characteristics are affected by number of factors such as; height of layers; type of the fabric and characteristics of the model (corners in the model, number of rounds and stops, number of markings and notches, numbers and characteristics of inner lines to be cut, cutting distance and number of patterns). All these factors affect cutting speed and determine cutting times in these computer aided cutting machines used in the study.

When cutting times of a circle and a square with same perimeter are examined, cutting time of the square is longer than the circle because the knife has to be taken out of the fabric layer three times due to the need for turnings at right angels. But there is no need for taking the knife out of the layer because there is no corner turning in cutting the circle shape and once cutting is started it continues without any pause, therefore, the speed will be higher and cutting time will be shorter. If number of marks, notches and hole marks is high, this will also slow down the system and prolong cutting time. When skirt models are examined with same concerns, number of stops will increase due to the number of darts, therefore, although the perimeter length is short cutting speed will slow down and cutting time will prolong.

The skirt and dress product groups are different from the lady's trousers and men's coat group in terms of structure. Taking the applied model criteria for the skirt and dress groups into consideration, when horizontally cut models are examined; secondary models are observed to have shorter perimeter of pieces. They are also cut in shorter cutting periods. The cutting time is closely related to not only the model but also to the structure and working characteristics of the computer aided cutting system used. Factors such as ; measurement of the marker, length of the cut, corner turnings, notches, holes and bents closely affect the cutting time.

In models with less pattern functions compared to skirt and dress such as, lady's trousers and men's coat, increment in the number of pieces and the perimeter of pieces prolong the cutting time.

Regression equalities for cutting time in lady's trousers and men's coat models are given below;

Lady's trousers cutting time (minute) = $5.005 + 7.16 \cdot 10^{-2} \times$ perimeter of pieces (m)
Men's coat cutting time (minute) = $3.438 + 8.99 \cdot 10^{-2} \times$ perimeter of pieces (m)

When number of pieces and sewing time is examined;

In all models subject to the study, increment in the number of pieces and the perimeter of pieces affect the sewing time. The increment in the number of pieces prolongs the length of the lines to be sewed. More lines to be sewed, means longer sewing time. Regression equalities for sewing times for skirt dress, lady's trousers and men's coat models are given below;

Skirt sewing time (min) = $0.276 + 0.282 \times$ perimeter of pieces (m)
Dress sewing time (min) = $7.724 + 0.197 \times$ perimeter of pieces (m)
Lady's sewing time (min) = $4.517 + 0.183 \times$ perimeter of pieces (m)
Men's coat sewing time (min) = $2.456 + 0.191 \times$ perimeter of pieces (m)

(Data for perimeter of pieces, cutting time, and sewing time are given in Appendix 37).

Following the investigations of all factors and their degrees of effect, it was observed that a characteristic of a model is one of the important factors that affect the productivity and time.

Time and productivity take the cost factor under its influence and divert it. Cost and profit calculations that only take into account the cost efficiency of the fabric are inappropriate.

Labor, a very important factor that affects the cost must be taken into consideration. The main factor affecting the labor cost is the time. In this respect, apparel manufacturers should handle both fabric and labor costs as a whole in cost calculations.

REFERENCES

Akgün, K., 1993, Bilgisayar destekli tasarımın üretim sürecindeki yeri, Bilgisayar Destekli Tasarım ve Ötesi Dergisi, Haziran, 4s.

Aldrich, W., 1992, CAD in Clothing and Textiles, A Collection of Expert Wiews, England, 179 s.

Anon, 1997, Frequently Asked CAD and Technology Realeted Questions, http: // www.techexchange. com / FAQ. html

Atılgan, T. ve Güner, M., 1999, Konfeksiyon ürünlerinin maliyet yapısı ve örnek uygulama, E.Ü. Tekstil ve Konfeksiyon Dergisi, Sayı: 6, 466 s.

Brackelsberg, P. and Marshall, R., 1990, Unit Method of Clotting Construction, Seventh Edition, United States of America, 280 p.

Carson, B.G., Bolz, A. and Young, H.H., 1972, Production Handbook, The Ronald Press Company, New York, 57 p.

Chase, R. and Aquiland, N.J., 1993, Production and Operations Management, Richard D. Irwin Inc., 194 p.

Chuter, A.J., 1990, Introduction to Clothing Production Management, Great Britain, 180 p.

Çavdarlı, S., 2002, İş ve insanın buluşma noktasında bir başkalaşıma doğru, İTKİB Hedef Dergisi, Sayı: 99, 112 s.

Eray, F. ve Çoruh, E., 2000, Hazır giyim endüstrisinde tasarım sürecinde kullanılan teknolojiler, E.Ü. Tekstil ve Konfeksiyon Dergisi, Sayı: 1-2, 50 s.

Erdoğan, M.Ç., 1991, İşçi elbisesi üretiminde ideal kumaş eninin saptanması, E.Ü. Tekstil ve Konfeksiyon Dergisi, Sayı:6, 621 s.

Erdoğan, M.Ç., 1992, Erkek takım elbisesi üretiminde ekose boyutlarının kumaş giderine etkisi, E.Ü. Tekstil ve Konfeksiyon Dergisi, Sayı: 3, 241 s.

Erdoğan, M.Ç., 1998, Modern kesim sistemleri, E.Ü. Tekstil ve Konfeksiyon Dergisi, Sayı:1 74 s.

Erdoğan, M.Ç., 1998, Modern kesim sistemleri, E.Ü. Tekstil ve Konfeksiyon Dergisi, Sayı:4, 288 s.

Groover, M.P. and Zimmers, E.W., 1984, CAD/CAM Computer Aided Designe and Manufacturing, USA, 489 p.

Güner, M., 2002, Dikim İşlemlerinin Birim Sürelerini Belirlemede Plan Zamanlar Yönteminin Uygulanması ve Bir Matematiksel Model Geliştirme, Doktora Tezi, E.Ü. Fen Bilimleri Enstitüsü, 109 s. (yayınlanmamış).

Heinrichs, T. and Bour, T., 1996, Konfeksiyon ve Hazır Giyimde Modern Bilgisayarlı Tasarım, Kalıp Çıkarma ve Serileme, Otomatik Pastal Kesme Sistemleri ve 2000'li Yıllara Bakış, VII. Uluslararası İzmir Tekstil ve Hazır Giyim Sempozyumu, İzmir.

İnceoğlu, Y., 2000, Giyside Temel Kalıp Çizimleri, İstanbul, 332 s.

Jerrigen, M.H. and Easterling, C.R., 1997, Fashion Merchandising and Marketing, Amazon Yayınları.

Kaptan, S., 1981, Bilimsel Araştırma Teknikleri ve İstatistik Yöntemleri, Tekışık Matbaası ve Rehber Yayınevi, Ankara, 66 s.

Karaca,   N.,   Regresyon   ve   Korelasyon   Analizi, http://www.borsacılar.com/eğitim/eğitim/regresyon_net.html.

Kabu, B., 1996, Üretim Yönetimi, İTÜ İşletme Fakültesi Yayını, No:1, İstanbul, 382 s.

Koç, Y., 1972, Sanayi İşletmelerinde Standart Maliyetler, Ankara Üniversitesi Basımevi, Ankara, 51s.

Konfeksiyon Teknolojisi Ansiklopedisi, 1995, Tekstil ve Konfeksiyon Araştırma Mrkz., Yayın No:55, Cilt: 2, İstanbul, 320 s.

Konfeksiyon Teknolojisi Ansiklopedisi, 1995, Tekstil ve Konfeksiyon Araştırma Merkezi, Yayın No:55, Cilt: 4, İstanbul, 331 s.

Markowitz, İ., 1988, Bilgisayar Kontrollü Kumaş Kesim Sisteminin Gelişimi ve Fiyat Performansı Değerlendirmesi, VIII Uluslararası İzmir Tekstil ve Hazır Giyim Sempozyumu, İzmir.

Obuchi, S. and Özerdağ, L., 1996, Dikimde Otomasyon Teknikleri, VII Uluslararası İzmir Tekstil ve Hazır Giyim Sempozyumu, İzmir.

Öndoğan, Z., 1997, Bilgisayar Destekli Tasarım, Kalıp, Model Uygulama ve Kesim Planı Hazırlama Sistemlerinin Hazır Giyim İşletmelerine Uyumu, Doktora Tezi, E.Ü. Fen Bilimleri Enstitüsü, İzmir. 119s (yayımlanmamış)

Öndoğan, Z., 1998, Hazır giyim sanayindeki talep değişimleri doğrultusunda tek kat kesimin değerlendirilmesi, E.Ü. Tekstil ve Konfeksiyon Dergisi, Sayı:6, 405 s.

Öndoğan, Z., 1999, Hazır giyim işletmelerinde kullanılan CAD sistemlerinin verimliliği, kullanıcıya ve işletmeye uygunluğu üzerine bir araştırma, E.Ü. Teksil ve Konfeksiyon Dergisi, Sayı:4, 338 s.

Özden, M., 1995, Konfeksiyon Giysi Üretiminde Kalıp Hazırlama, Serilendirme, Kesim Planı Yerleştirme ve Çizim İşlemlerinde Kullanılacak Komple Bir CAD Programının Geliştirilmesi, Doktora Tezi, E.Ü. Fen Bilimleri Enstitüsü, 132 s (yayımlanmamış).

Özden, M., ve Göktogan, A.H., 1996, Türkiye'de Doğan ve Bugünün Globalleşen Dünyasında Değişim Yaratacak Konfeksiyon CAD Teknolojisi: Konsan CAD, VII Uluslararası İzmir Tekstil ve Hazır Giyim Sempozyumu, İzmir.

Püskülcü, H. ve İkiz, F., 1986, İstatistiğe Giriş, E.Ü.Mühendislik Fakültesi Ders Kitapları, No:1, İzmir, 333 s.

REFA, 1988, İş Etüdü Yöntem Bilgisi, Kitap 2 Veri Saptama, İş Etüdü ve İşletme Organizasyonu Birliği, 447 s.

REFA, 1988, İş Etüdü Yöntem Bilgisi, Kitap 3 Maliyet Muhasebesi, İş Düzenlemesi, İş Etüdü ve İşletme Organizasyonu Birliği, 387 s.

Taylor, P., 1995, Giyim Endüstrisinde Bilgisayar (çev:A. Ulucan) Gaye Filmcilik Matbaacılık San.Tic.A.Ş., Ankara, 220 s.

Taylor, P. and Shoben, M.M., 1995, Giyim Sanayi İçin Serileştirme Kural ve Uygulama (çev: H.Bal, R.Akay), Gaye Filmcilik Matbaacılık San.Tic. A.Ş., Ankara, 280 s.

APPENDIX 1. Women size table

SIZE	Ow	Tw	Hw	Sp	Rb	Ad	Ab	Sb	HI	VI	Sh	Ht	Bt	Ah	Oaw	AI	Sl	Rol	Sz	Kn
36	42	31.75	45	6.2	16.5	8.5	7.5	12	42	44.5	20.9	21	25.5	15.6	26	42	104	60	27	23
38	44	34	47	6.4	17	9	8.5	12.2	42	45	20.7	21	26.1	15.8	27.1	42	104	60	27	23.5
40	46	36.25	49	6.6	17.5	9.5	9.5	12.4	42	45.5	20.5	21	26.7	16	28.2	42	104	60	27	24
42	48	38.5	51	6.8	18	10	10.5	12.6	42	46	20.3	21	20.3	16.2	29.3	42	104	60	27	24.5
44	50	40.75	53	7	18.5	10.5	11.5	12.8	42	46.5	20.1	21	20.9	16.4	30.4	42	104	60	27	25
46	52	43	55	7.2	19	11	12.5	13	42	47	19.9	21	28.5	16.6	31.5	42	104	60	27	25.5
48	55	46	57	7.4	19.5	11.9	14.2	13.4	42	47.7	19.5	21	25.6	17	32.9	42	104	60	27	26

APPENDIX 2. Men size table

SIZE	Bu	Rb	Ad	Bb	RI	VI	Sh	Sp	Lg	Sb
44	88	18.5	11	18.5	42	40	22.2	7.4	80	15
46	92	19	11.5	19	43	41	22.2	7.6	80	15.5
48	96	19.5	12	19.5	44	42	22.2	7.8	80	16
50	100	20	12.5	20	45	43	22.2	8	80	16.5
52	104	20.5	13	20.5	46	44	22.2	8.2	80	17
54	108	21	13.5	21	46.5	45	22.2	8.4	80	17.5

APPENDIX 3. Classic skirt

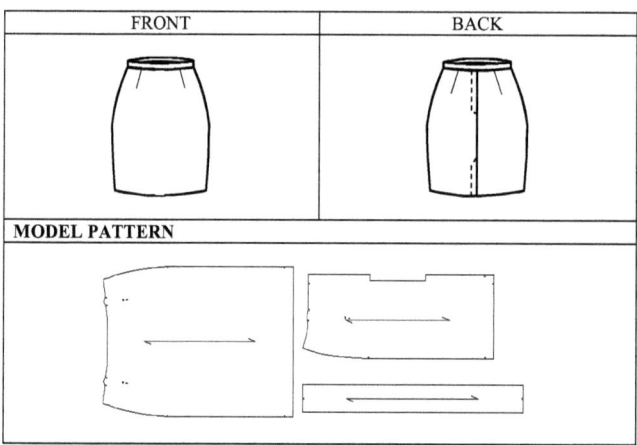

APPENDIX 4. Flared skirt

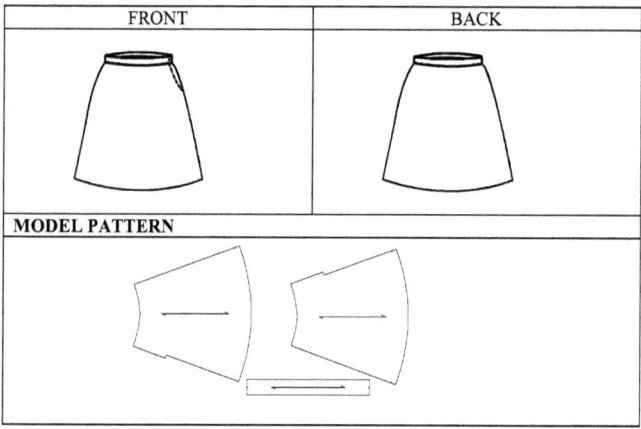

APPENDIX 5. Horizontally cut model 1 in skirt

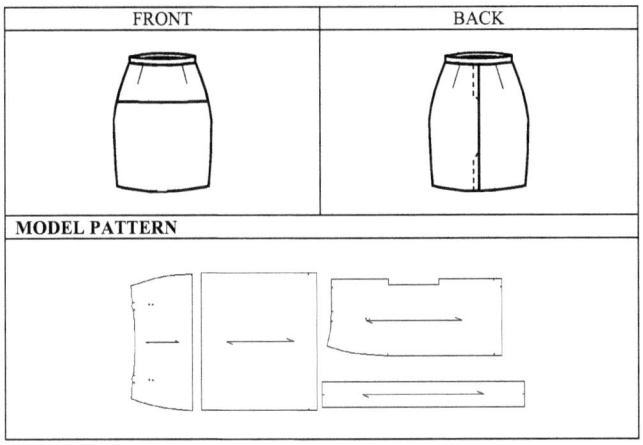

APPENDIX 6. Horizontally cut model 2 in skirt

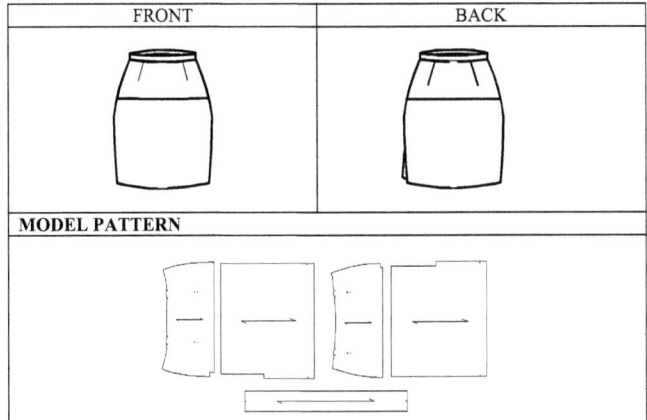

APPENDIX 7. Vertically cut model 1 in skirt

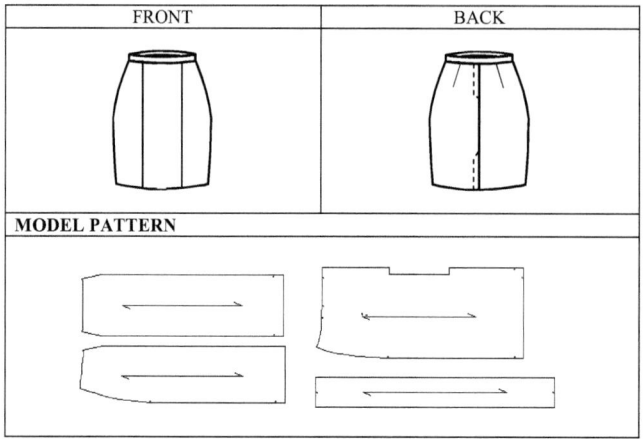

APPENDIX 8. Vertically cut model 2 in skirt

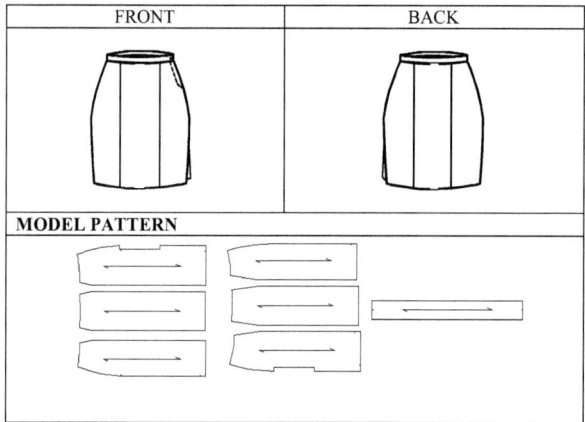

APPENDIX 9. Both vertically and horizontally cut model 1 in skirt

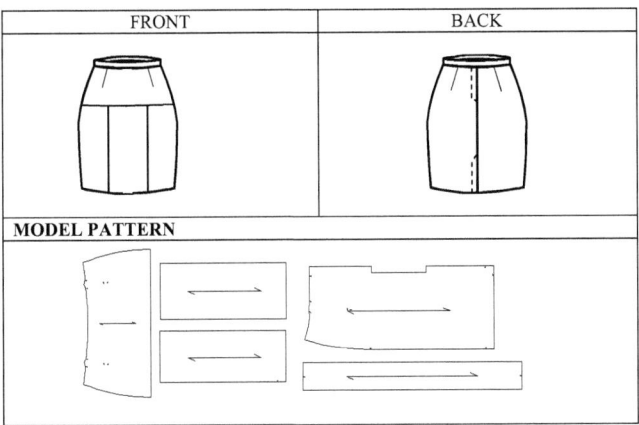

APPENDIX 10. Both vertically and horizontally cut model 2 in skirt

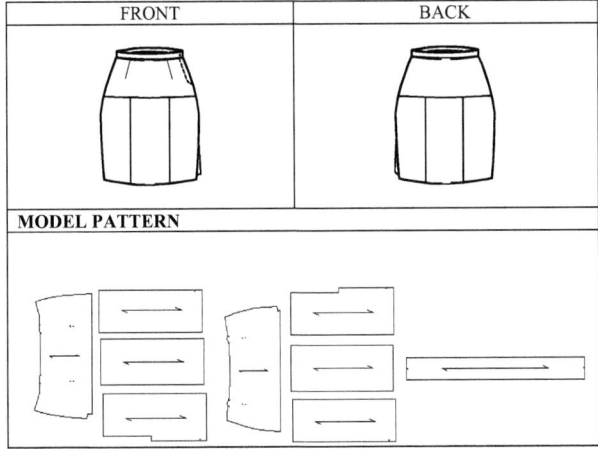

APPENDIX 11. Classic lady's trousers

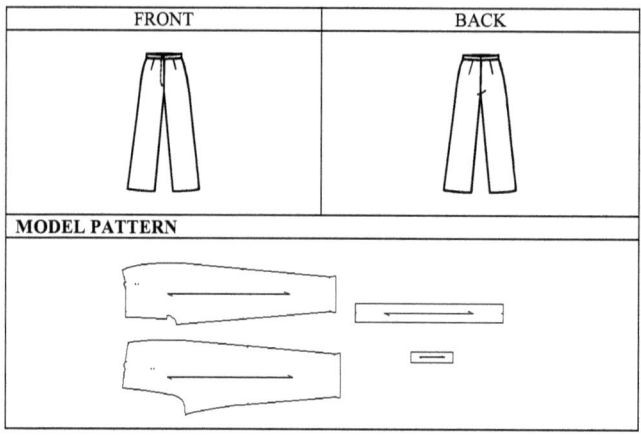

APPENDIX 12. Flared lady's trousers

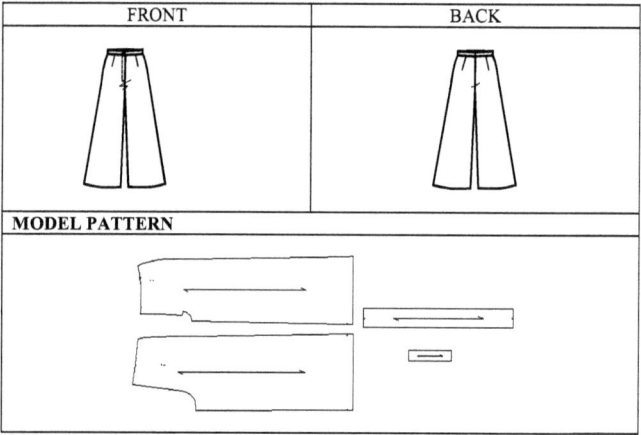

APPENDIX 13. Horizontally cut model 1 in lady's trousers

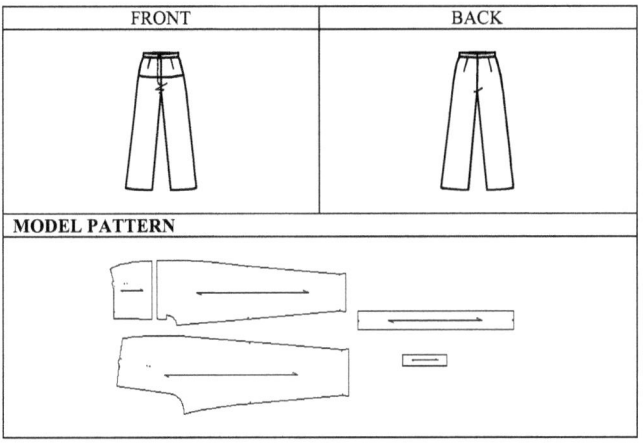

FRONT	BACK
MODEL PATTERN	

APPENDIX 14. Horizontally cut model 2 in lady's trousers

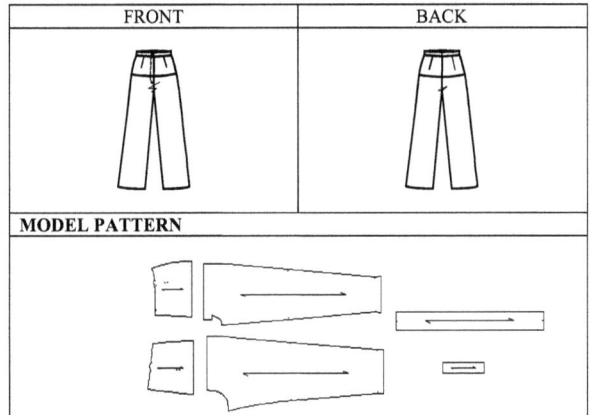

FRONT	BACK
MODEL PATTERN	

APPENDIX 15. Vertically cut model 1 in lady's trousers

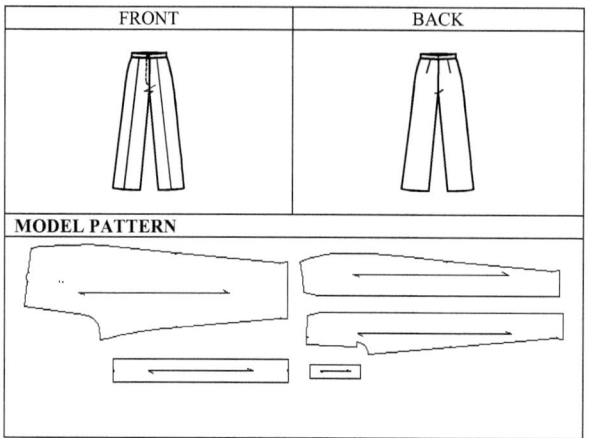

APPENDIX 16. Vertically cut model 2 in lady's trousers

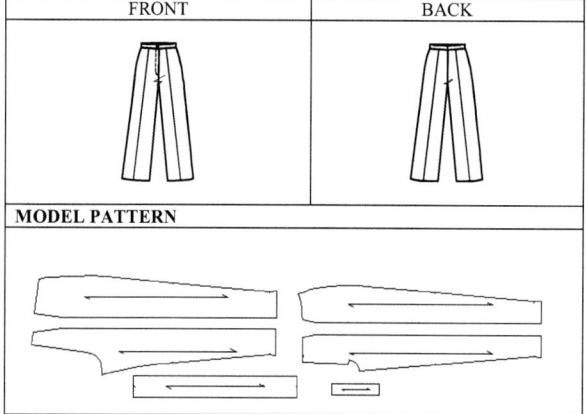

APPENDIX 17. Both vertically and horizontally cut model 1 in lady's trousers

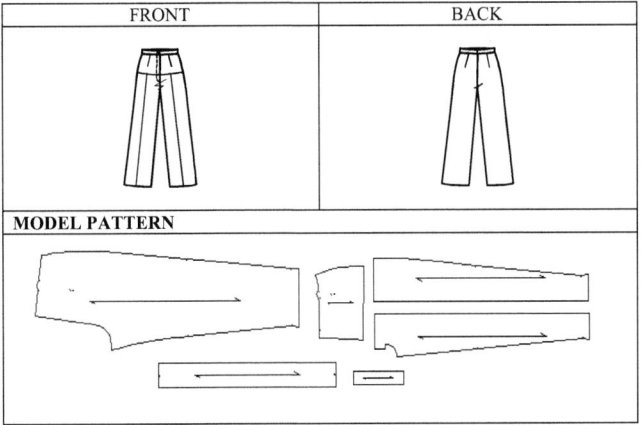

APPENDIX 18. Both vertically and horizontally cut model 2 in lady's trousers

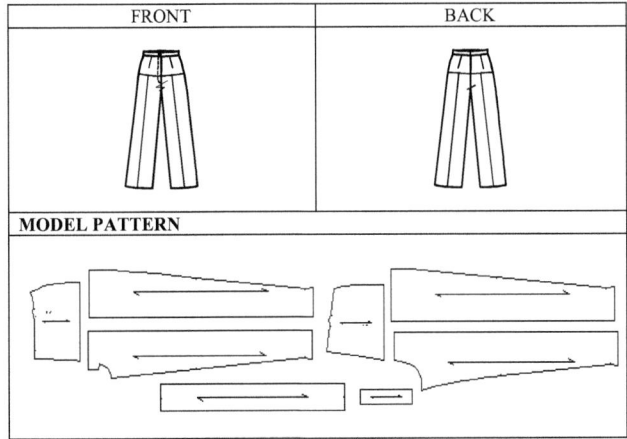

APPENDIX 19. Classic dress

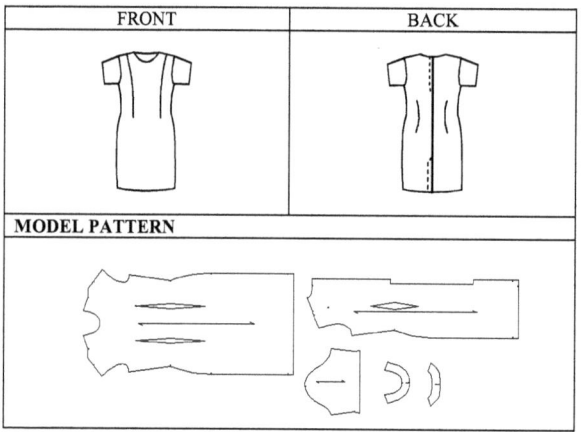

FRONT	BACK
MODEL PATTERN	

APPENDIX 20. Flared dress

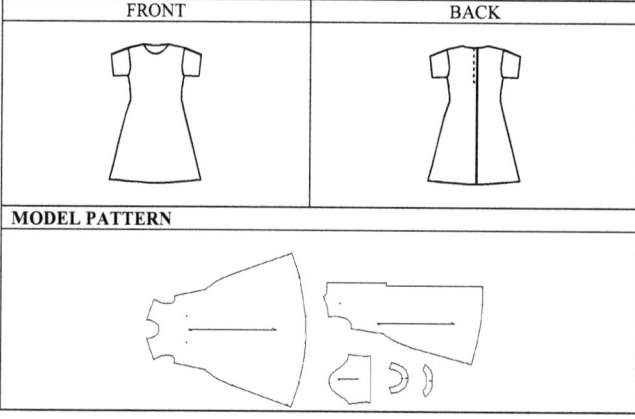

FRONT	BACK
MODEL PATTERN	

APPENDIX 21. Horizontally cut model 1 in dress

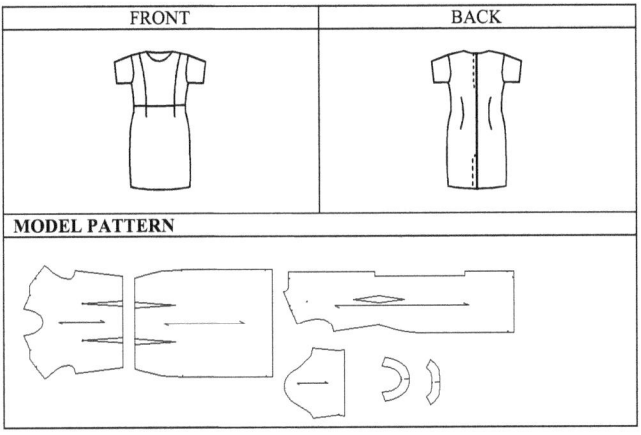

APPENDIX 22. Horizontally cut model 2 in dress

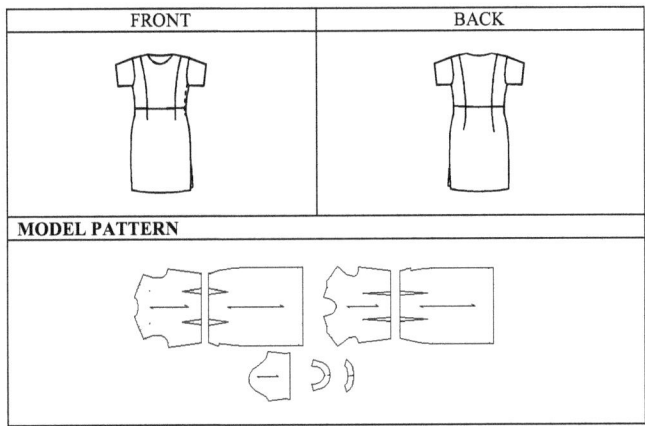

APPENDIX 23. Vertically cut model 1 in dress

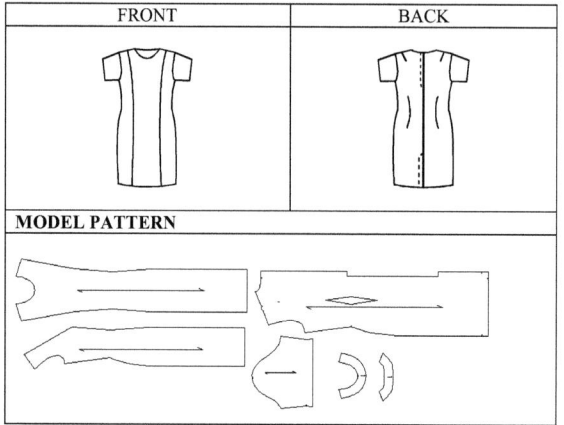

APPENDIX 24. Vertically cut model 2 in dress

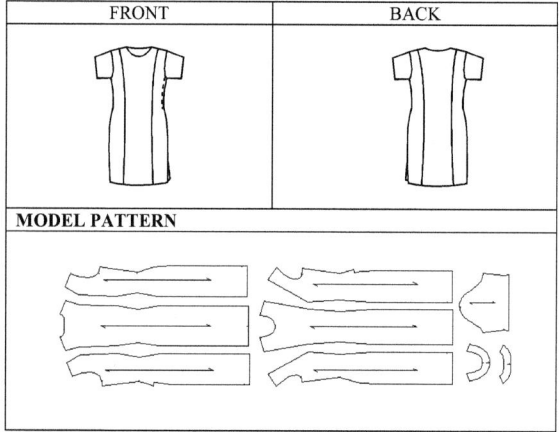

APPENDIX 25. Both vertically and horizontally cut model 1 in dress

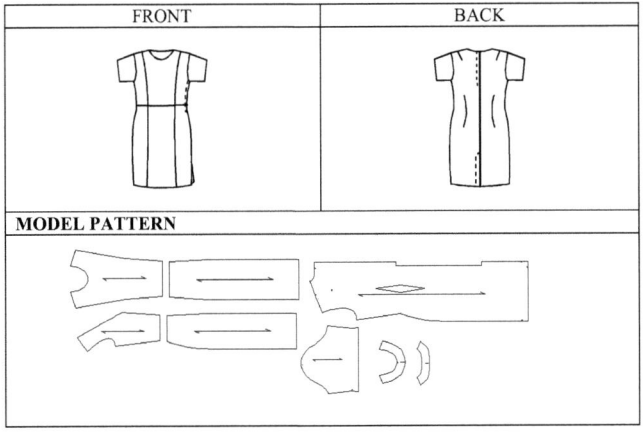

APPENDIX 26. Both vertically and horizontally cut model 2 in dress

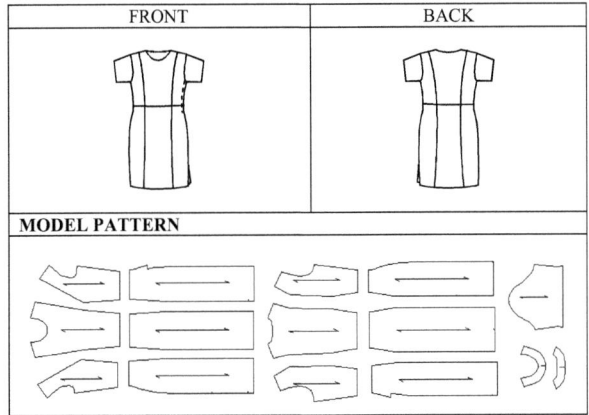

APPENDIX 27. Classic men's coat

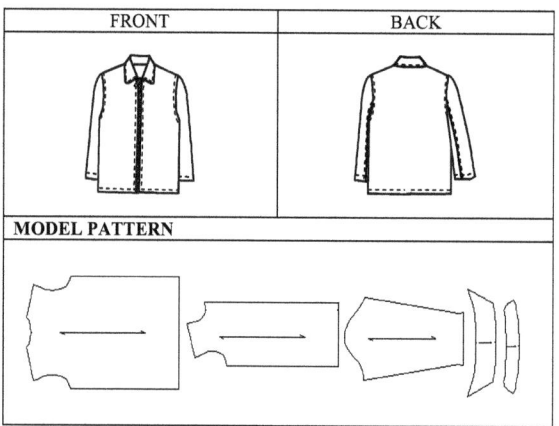

APPENDIX 28. Horizontally cut model 1 in men's coat

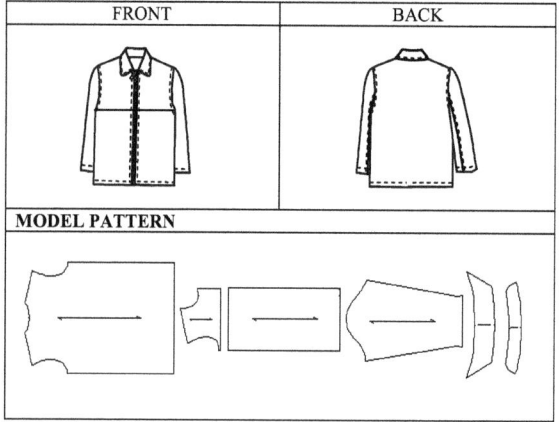

APPENDIX 29. Horizontally cut model 2 in men's coat

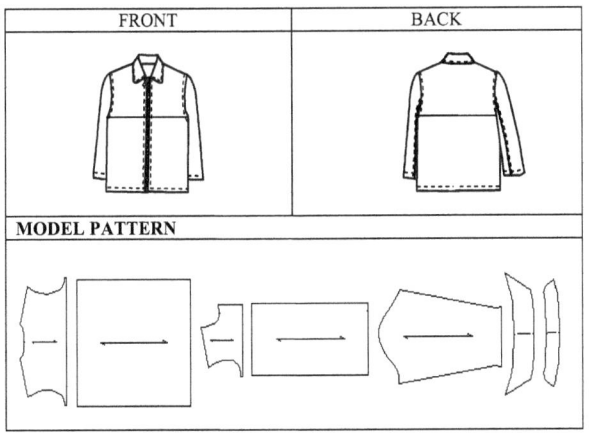

APPENDIX 30. Vertically cut model 1 in men's coat

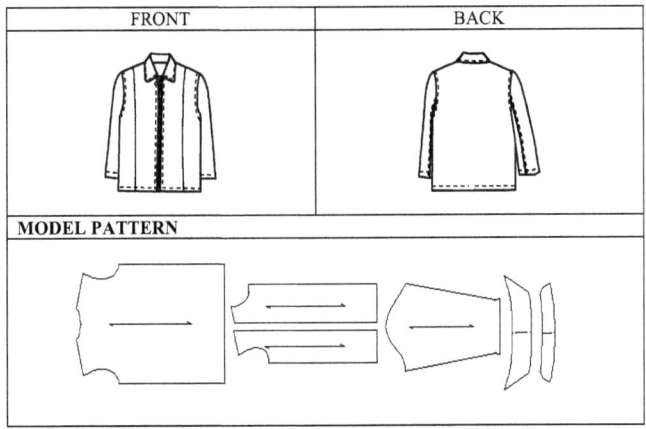

APPENDIX 31. Vertically cut model 2 in men's coat

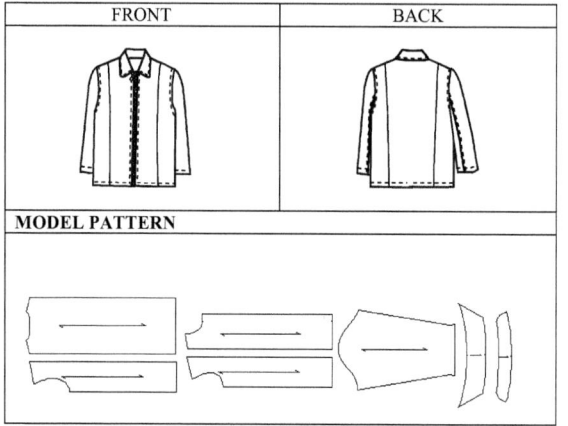

APPENDIX 32. Both vertically and horizontally cut model 1 in men's coat

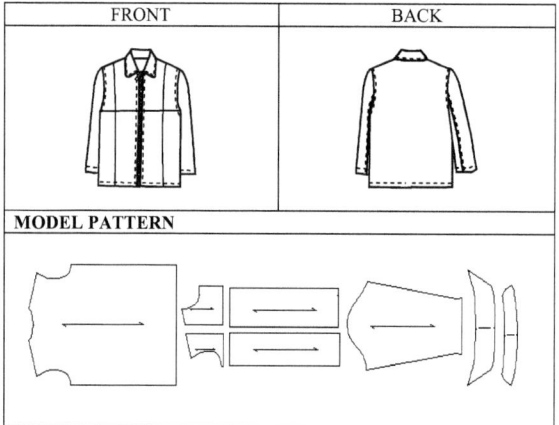

APPENDIX 33. Both vertically and horizontally cut model 2 in men's coat

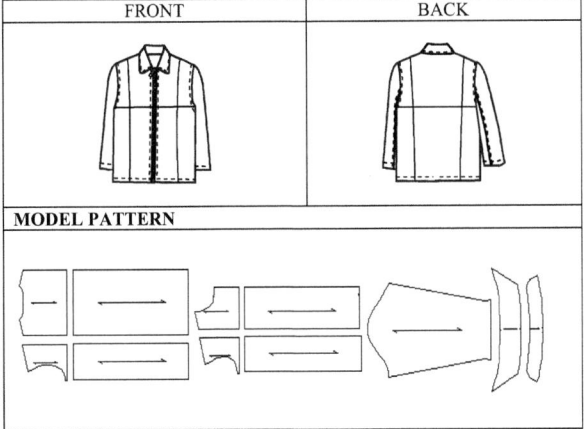

FRONT	BACK

MODEL PATTERN

APPENDIX 34. Research findings accoding to fabric usage proportion, total number of pieces on the marker, area of pieces and perimeter of pieces.

SKIRT MODELS	FABRIC USAGE PROPORTION (%)	TOTAL NUMBER OF PIECES	AREA OF PIECES (m²)	PERIMETER OF PIECES (m)
CLASİC	77,1	16	2,87	30,7
FLARED	83,3	12	3,49	27,5
HORIZANTALLY CUT MODEL 1	90,4	20	2,91	34,9
HORIZANTALLY CUT MODEL 2	86,4	20	2,9	33,7
VERTICALLY CUT MODEL 1	79,1	24	2,97	40,9
VERTICALLY CUT MODEL 2	80,2	28	3	45,7
BOTH VERTICALLY AND HORIZANTALLY CUT MODEL 1	92,3	28	2,98	42
BOTH VERTICALLY AND HORIZANTALLY CUT MODEL 2	92,8	36	3,04	48,1

LADY'S TROUSERS MODELS				
CLASİC	78,9	24	4,79	57,6
FLARED	90,5	24	5,87	54
HORIZANTALLY CUT MODEL 1	84,6	32	4,83	56,7
HORIZANTALLY CUT MODEL 2	88	40	4,87	61,2
VERTICALLY CUT MODEL 1	80,6	32	4,95	69
VERTICALLY CUT MODEL 2	84,4	40	5,11	86,7
BOTH VERTICALLY AND HORIZANTALLY CUT MODEL 1	85,9	40	4,98	71,4
BOTH VERTICALLY AND HORIZANTALLY CUT MODEL 2	92,7	56	5,16	90,6

DRESS MODELS				
CLASİC	82,2	28	4,91	48,3
FLARED	84,6	28	6,61	51,2
HORIZANTALLY CUT MODEL 1	85,2	32	4,95	51,9
HORIZANTALLY CUT MODEL 2	87,6	32	4,85	47
VERTICALLY CUT MODEL 1	83,3	36	4,96	64,8
VERTICALLY CUT MODEL 2	81,6	40	4,95	73
BOTH VERTICALLY AND HORIZANTALLY CUT MODEL 1	83,1	48	5	68,7
BOTH VERTICALLY AND HORIZANTALLY CUT MODEL 2	85,6	64	5,03	81

MEN'S COAT MODELS				
CLASİC	85,9	36	6,1	62,9
HORIZANTALLY CUT MODEL 1	86,4	44	6,15	68,6
HORIZANTALLY CUT MODEL 2	87,1	48	6,2	73,4
VERTICALLY CUT MODEL 1	87,6	44	6,23	76,2
VERTICALLY CUT MODEL 2	89,3	52	6,36	89,6
BOTH VERTICALLY AND HORIZANTALLY CUT MODEL 1	88,9	60	6,29	82,3
BOTH VERTICALLY AND HORIZANTALLY CUT MODEL 2	89,5	80	6,47	101

47

APPENDIX 35. Research findings according to the number of pieces and cutting time in marker plan.

SKIRT MODELS	CUTTING TIME (minute)	NUMBER OF PIECES
CLASİC	6,66	16
FLARED	3,83	12
HORIZANTALLY CUT MODEL 1	6,96	20
HORIZANTALLY CUT MODEL 2	5,19	20
VERTICALLY CUT MODEL 1	7,64	24
VERTICALLY CUT MODEL 2	6,18	28
BOTH VERTICALLY AND HORIZANTALLY CUT MODEL 1	7,35	28
BOTH VERTICALLY AND HORIZANTALLY CUT MODEL 2	6,66	36

LADY'S TROUSERS MODELS		
CLASİC	9,91	24
FLARED	6,96	24
HORIZANTALLY CUT MODEL 1	9,61	32
HORIZANTALLY CUT MODEL 2	10,37	40
VERTICALLY CUT MODEL 1	9,71	32
VERTICALLY CUT MODEL 2	10,41	40
BOTH VERTICALLY AND HORIZANTALLY CUT MODEL 1	10,4	40
BOTH VERTICALLY AND HORIZANTALLY CUT MODEL 2	11,87	56

DRESS MODELS		
CLASİC	10,6	28
FLARED	11,07	28
HORIZANTALLY CUT MODEL 1	10,67	32
HORIZANTALLY CUT MODEL 2	8,92	32
VERTICALLY CUT MODEL 1	9,79	36
VERTICALLY CUT MODEL 2	10,68	40
BOTH VERTICALLY AND HORIZANTALLY CUT MODEL 1	10,89	48
BOTH VERTICALLY AND HORIZANTALLY CUT MODEL 2	12,25	64

MEN'S COAT MODELS		
CLASİC	8,81	36
HORIZANTALLY CUT MODEL 1	9,82	44
HORIZANTALLY CUT MODEL 2	10,4	48
VERTICALLY CUT MODEL 1	10,09	44
VERTICALLY CUT MODEL 2	10,19	52
BOTH VERTICALLY AND HORIZANTALLY CUT MODEL 1	11,55	60
BOTH VERTICALLY AND HORIZANTALLY CUT MODEL 2	13	80

APPENDIX 36. Research findings according to the number of pieces and sewing time

SKIRT MODELS	SEWING TIME (minute)	NUMBER OF PIECES
CLASİC	9,539	4
FLARED	6,792	3
HORIZANTALLY CUT MODEL 1	10,628	5
HORIZANTALLY CUT MODEL 2	10,567	5
VERTICALLY CUT MODEL 1	11,351	6
VERTICALLY CUT MODEL 2	11,991	7
BOTH VERTICALLY AND HORIZANTALLY CUT MODEL 1	12,545	7
BOTH VERTICALLY AND HORIZANTALLY CUT MODEL 2	14,333	9

LADY'S TROUSERS MODELS		
CLASİC	13,715	6
FLARED	14,931	6
HORIZANTALLY CUT MODEL 1	15,195	8
HORIZANTALLY CUT MODEL 2	16,689	10
VERTICALLY CUT MODEL 1	16,527	8
VERTICALLY CUT MODEL 2	19,463	10
BOTH VERTICALLY AND HORIZANTALLY CUT MODEL 1	17,907	10
BOTH VERTICALLY AND HORIZANTALLY CUT MODEL 2	22,013	14

DRESS MODELS		
CLASİC	17,9	7
FLARED	14,603	7
HORIZANTALLY CUT MODEL 1	18,988	8
HORIZANTALLY CUT MODEL 2	19,06	8
VERTICALLY CUT MODEL 1	19,264	9
VERTICALLY CUT MODEL 2	20,014	10
BOTH VERTICALLY AND HORIZANTALLY CUT MODEL 1	22,252	12
BOTH VERTICALLY AND HORIZANTALLY CUT MODEL 2	25,576	16

MEN'S COAT MODELS		
CLASİC	14,733	9
HORIZANTALLY CUT MODEL 1	16,141	11
HORIZANTALLY CUT MODEL 2	15,619	12
VERTICALLY CUT MODEL 1	16,849	11
VERTICALLY CUT MODEL 2	19,061	13
BOTH VERTICALLY AND HORIZANTALLY CUT MODEL 1	18,585	15
BOTH VERTICALLY AND HORIZANTALLY CUT MODEL 2	22,174	20

APPENDIX 37. Research findings according to the perimeter of pieces, cutting time, and sewing time

SKIRT MODELS	CUTTING TIME (minute)	SEWING TIME (minute)	PERIMETER OF PIECES (m)
CLASİC	6,66	9,539	30,7
FLARED	3,83	6,792	27,5
HORIZANTALLY CUT MODEL 1	6,96	10,628	34,9
HORIZANTALLY CUT MODEL 2	5,19	10,567	33,7
VERTICALLY CUT MODEL 1	7,64	11,351	40,9
VERTICALLY CUT MODEL 2	6,18	11,991	45,7
BOTH VERTICALLY AND HORIZANTALLY CUT MODEL 1	7,35	12,545	42
BOTH VERTICALLY AND HORIZANTALLY CUT MODEL 2	6,66	14,333	48,1

LADY'S TROUSERS MODELS			
CLASİC	9,91	13,715	57,6
FLARED	6,96	14,931	54
HORIZANTALLY CUT MODEL 1	9,61	15,195	56,7
HORIZANTALLY CUT MODEL 2	10,37	16,689	61,2
VERTICALLY CUT MODEL 1	9,71	16,527	69
VERTICALLY CUT MODEL 2	10,41	19,463	86,7
BOTH VERTICALLY AND HORIZANTALLY CUT MODEL 1	10,4	17,907	71,4
BOTH VERTICALLY AND HORIZANTALLY CUT MODEL 2	11,87	22,013	90,6

DRESS MODELS			
CLASİC	10,6	17,9	48,3
FLARED	11,07	14,603	51,2
HORIZANTALLY CUT MODEL 1	10,67	18,988	51,9
HORIZANTALLY CUT MODEL 2	8,92	19,06	47
VERTICALLY CUT MODEL 1	9,79	19,264	64,8
VERTICALLY CUT MODEL 2	10,68	20,014	73
BOTH VERTICALLY AND HORIZANTALLY CUT MODEL 1	10,89	22,252	68,7
BOTH VERTICALLY AND HORIZANTALLY CUT MODEL 2	12,25	25,576	81

MEN'S COAT MODELS			
CLASİC	8,81	14,733	62,9
HORIZANTALLY CUT MODEL 1	9,82	16,141	68,6
HORIZANTALLY CUT MODEL 2	10,4	15,619	73,4
VERTICALLY CUT MODEL 1	10,09	16,849	76,2
VERTICALLY CUT MODEL 2	10,19	19,061	89,6
BOTH VERTICALLY AND HORIZANTALLY CUT MODEL 1	11,55	18,585	82,3
BOTH VERTICALLY AND HORIZANTALLY CUT MODEL 2	13	22,174	101

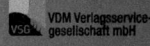

Printed by Books on Demand GmbH, Norderstedt / Germany